Huntington Richards

A Concluding Report on the Anatomy of the Elephant's Ear

Huntington Richards

A Concluding Report on the Anatomy of the Elephant's Ear

ISBN/EAN: 9783337866099

Printed in Europe, USA, Canada, Australia, Japan

Cover: Foto ©berggeist007 / pixelio.de

More available books at **www.hansebooks.com**

A CONCLUDING REPORT ON THE

ANATOMY OF THE ELEPHANT'S EAR,

BY

HUNTINGTON RICHARDS, M. D.,

––––––––

[Reprinted from The Transactions of the AMERICAN OTOLOGICAL SOCIETY, 1891.]

A CONCLUDING REPORT ON THE ANATOMY OF THE ELEPHANT'S EAR.

By Huntington Richards, M. D.

In a previous paper, discussing and attempting to illustrate and explain the anatomy of the elephant's ear, I deferred a more detailed account of the form, relations, and connections of the ossicles, and also fuller discussion of the labyrinthine structures, "until the issue of still another paper" (vide Transactions of the American Otological Society, Vol. 4, Part 4, p. 599). Circumstances unforeseen, and unexpected by me at the time of writing and presenting that earlier paper, have rendered it difficult for me to study those especial points, and at the very outset of this present report — the last that I purpose offering to the Society on this subject, and, therefore, a concluding report, as far as the writer is concerned — I must apologize for its almost total lack of additional information regarding these points.

The earlier dissection of the specimen, made by Dr. Buck in the preparation of his, the first report, necessarily opened the cavity of the vestibule and cut across two of the semicircular canals. The necessity for exposing these parts at that time arose from the fact that, owing to the narrowness of the tympanic vault and the thickness of its bony roof, it could not be seen without extensive removal of the pars petrosa (vide Dr. Buck's paper, Transactions, Vol. 4, Part 4, p. 586, at top). The vestibule, as now exposed in the specimen before me, is an irregular cup-shaped cavity, six millimeters deep, and measuring across its mouth nine millimeters in an extero-internal and seven millimeters in an antero-posterior direction. At the entrances of the semicircular canals and at the entrance of the cochlear canals, the membranous structures are still present; elsewhere throughout its cavity they have been removed. The footplate of the stirrup, measuring about five by four millimeters, occupies most of its outer wall. The cochlea is still intact and lies unexamined within its bony envelope. I shall leave its proper investigation in the case of this specimen to some other observer. To him also I leave a more detailed discussion of the vestibule cavity and of the one intact and two mutilated semicircular canals.

To obtain a clearer view of the ossicles and their relations, and to facilitate search for the opening of the fenestra rotunda, I separated the specimen previously described by myself (article "A further report, etc."), into an outer and an inner fragment. This separation was effected by sawing through the anterior inferior wall of the Fallopian canal, thus opening into the back of the tympanic vault, and through the bony structures lying behind and below the tympanum, and then, — following the line of accidental separation of the thick membranous floor of the vault (see preceding paper, p. 603), — by divulsion, along the line of suture running between the periotic bone at the promontory and the squamosal bone. The outer fragment carried the membrana tympani, malleus, and incus; the inner retained the stirrup. This inner fragment was, by accidental divulsion, separated into a larger and a smaller portion, this second line of separation following the occipito-squamoso-periotic suture (vide Fig. 2 in my previous paper and its discussion, at top of page 595, and also Fig. 4 of the present paper) and the two lines of accidental fracture across the bone that intervened between this suture and the antrum cavity and across the bony septum between this cavity and the Fallopian canal. This separation or divulsion exposed the cavity containing the body of the stapedius muscle,—that cavity lying between the upper part of the Fallopian canal and the antrum. The split through the bone which extended into this cavity is plainly visible in Fig. 2 of the former and in Fig. 4 of this present paper. The forcible divulsion of these three fragments, chiefly, as I recall it, that of the latter two, caused an accidental fracture of both crura stapedis, the footplate remaining firmly fixed in the fenestra ovalis, the crura and head remaining attached to the stapedius muscle. An interesting fact is demonstrated by this disruption of the specimen, viz., that there is in the elephant no complete bony division between the cavity containing the belly of the stapedius and the general cavity of the tympanum, an hiatus in the wall of the former, and one of considerable size, existing on its inner aspect, and consequently on the outer wall of the secondary tympanic cavity. This hiatus, this opening of communication (so far as true bony separation is concerned), lies behind the pars pyramidalis, quite outside of the tympanum proper and at a point in the outer wall of the secondary tympanic cavity, opposite to and on a level with the inner opening of a canal shown in Fig. 3 of the preceding paper and described (see top of page 603, in report for 1891) as that carrying a branch of the stylo-mastoid artery. Further description of the stapedius muscle I leave to the writer of a still further report upon this most interesting anatomical specimen. Of the fenestra rotunda I can discover no trace. A determination of the mooted question of its presence or absence in the elephant (see Dr. Buck's paper, footnote on pp. 582, 583) can in my opinion be best obtained by dissection of the cochlea and careful tracing of the scala-tympani to its point of termination. The "hooded opening or niche," lying posterior to the "analogue of the promontory," and therefore in the outer wall of the upper portion of the secondary tympanic cavity, which Dr. Buck suggested as being, but did not believe to be, the niche of the fenestra rotunda, is plainly shown by my further dissection to lie wholly above and behind the vestibule and, consequently, to be in close relation with the semicircular canals and far removed from the cochlea. This hooded niche lies in close relation, also, to the cavity containing the belly of the stapedius. It is the posterior end of a narrow groove, about 8 millimeters long, 2½ millimeters wide, and 2½ millimeters long, which extends in an antero-posterior direction along the back surface of the promontory of the pars petrosa, and which, at its forward end (wholly hidden from view until this final disruptive dissection of the specimen), is separated by membranous structures only from the cavity of the vestibule.

Before proceeding to comment upon the four illustrations presented in this, my own concluding paper, I will make bold to steal a little of the thunder of my successor in calling attention to a curious formation noticeable in the long process of the anvil. At least in the specimen before me, this process is not a solid shaft of bone; its lower (or, rather, posterior) half, where it curves inward for articulation with the head of the stapes, shows a tendency toward the tubular form, the inner wall of the tube or hollow cylinder being absent just above (or, rather, in front of) the articular tip, so that an oval opening with smooth rounded edges is presented at that point of its surface constituting an hiatus in the inner wall of the bony shaft. A like tendency is noticeable in the semi-cylindrical form of the crura stapedis, in apparent obedience to a general rule of formation observable in all the bone structure of the elephant's cranium, viz., such economy of solid bone tissue as to result in a maximum amount of strength and surface and a minimum of weight. The prevailing feature of honey-comb or cancellar structure is throughout most singularly noticeable and characteristic.

The main object of this paper is the presentation of four illustrations, made by photo-engraving from photographs taken by myself, which I trust may serve as supplements to those already published in Dr. Buck's paper and in my own former paper. A detailed comment upon these new illustrations is unnecessary, hence I shall merely call attention to points shown in them better than they were shown in previously presented views of the specimen, and to things shown from a slightly different point of view, trusting that this new explanatory text and these new pictures may serve to give to such readers as will be at the pains to refer to former views and former explanations a more accurate notion of the peculiar features of the complicated anatomical specimen now so long under discussion.

Fig. 1. In this picture, the specimen is looked at from a point of view similar to, but not quite identical with that adopted in the case of Fig. 3 of my previous report. The top of the specimen is not seen in the illustration; its position is to the left of and beyond the margin of the engraving.

In Fig. 3 of the former paper, the top of the illustration corresponded with the upper end of the specimen; here, this is not the case but the bone is seen lying over to the left, on its forward and outer edge. I present this illustration in order to give a view of the curious drumstick-like osteophyte described by Dr. Buck (loc. cit., p. 579) which will do fuller justice to the subject than did the photograph made for Fig. 3 of his paper, and which will also show the margins of the neighboring crypts better than they were shown in Fig. 3 of my own former paper. The drumstick osteophyte is seen in the right-hand half of the present illustration, close to a copper wire which was bound around the specimen subsequent to the accident described

in the last paper. The crypt mouths lie behind!(in the
illustration *above*) the central portion of the specimen. This
central portion is the remaining part of the pars petrosa :
in it (about three-quarters of an inch above the lower bor-
der of the illustration) is plainly to be seen the now exposed
cavity of the vestibule. Almost coincident with the lower
margin of the illustration may be seen a cross section of one
of the semicircular canals. Any careful reader of this

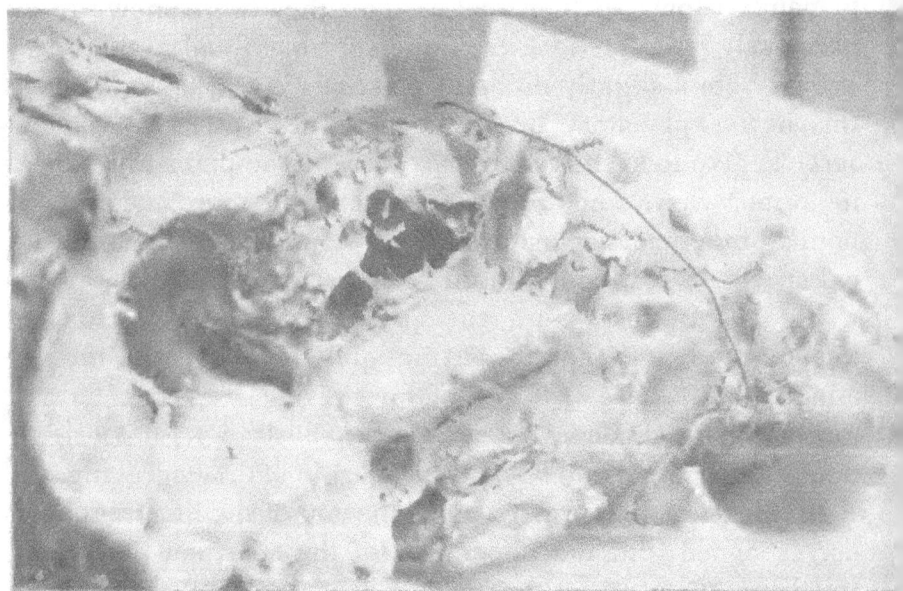

FIG. I.

paper and of either one or both of those which have preceded
it, will at once recognize the large cavity seen towards the
left-hand margin of this Fig. 1, as being the upper portion
of the secondary or subsidiary tympanic cavity. In the
specimen, the parts are about one-fifth larger than they are
shown in the illustration.

Fig. 2. This third attempt to show the roof of the tympanum proper is here introduced as supplementary to Fig.

2 of Dr. Buck's paper and to Fig. 4 of my former paper. Be it remembered that in this view the observer is looking directly upwards in a vertical line to the membranous roof of the tympanum and sees the umbo of the membrana tympani projecting upwards like a tent and the handle of the malleus very much foreshortened. The smooth, rounded, lower surface of the pars petrosa, the promontory or analogue of the promontory, occupies the centre of the illustration. If the engraver succeeds in clearly reproducing the photograph, the course of the chorda tympani

FIG. 2.

nerve will be plainly seen to the left, and the mouths of several crypts opening into the subsidiary tympanum will be plainly seen to the right, of this promontory. To one of these crypt mouths, that one which shows most distinctly in the figure and which is triangular in outline, as there seen, I desire to call special attention. At the bottom of this crypt is situated the hiatus in the outer bone wall of the subsidiary tympanum, where membranous tissue alone intervenes between this cavity and that containing the belly of the stapedius muscle. In the illustration, the parts are shown a trifle larger than life size.

Fig. 3. This illustration, engraved from a photograph so taken as to magnify the object, shows about the same points discussed under the head of Fig. 1; but the parts are here seen about one and one-half times as large as in nature. The point of view is slightly different from that chosen for Fig. 1, and the reader will observe that

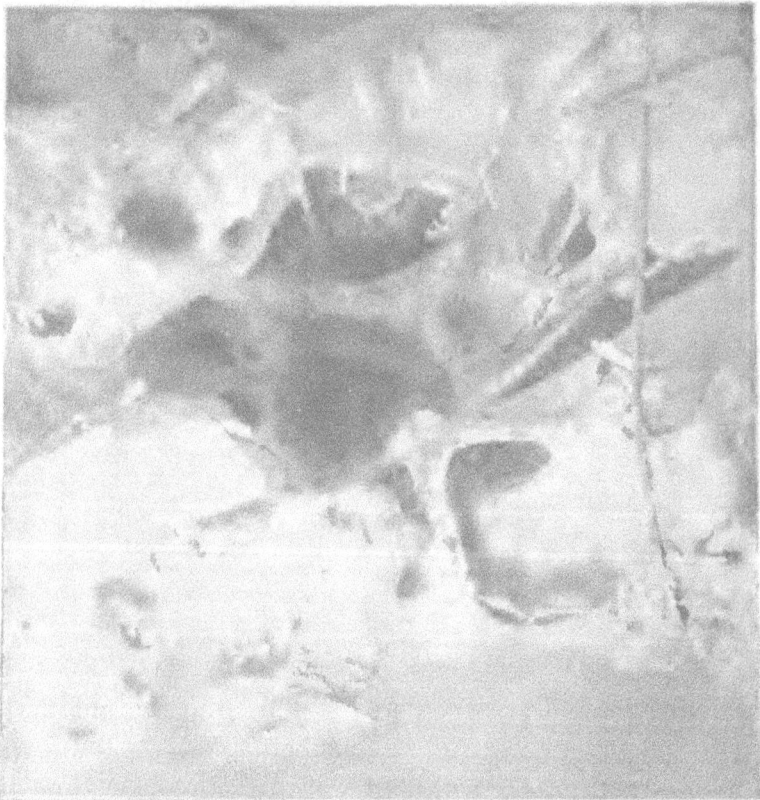

Fig. 3.

the wire, which in Fig. 1 crosses the picture above the shank of the drumstick osteophyte, here shows as crossing the knob-like extremity of that bony spur.

Fig. 4. In this illustration, the parts are again shown magnified to nearly twice their natural size. If the engraving succeeds in clearly reproducing the photograph, the picture will give a good view of the hammer, anvil, and stirrup; of the tendon of the stapedius; of the antrum; of the sutures described in my former paper; and of the accidental bone crack leading from antrum to tympanum and (as explained in the early part of this present report)

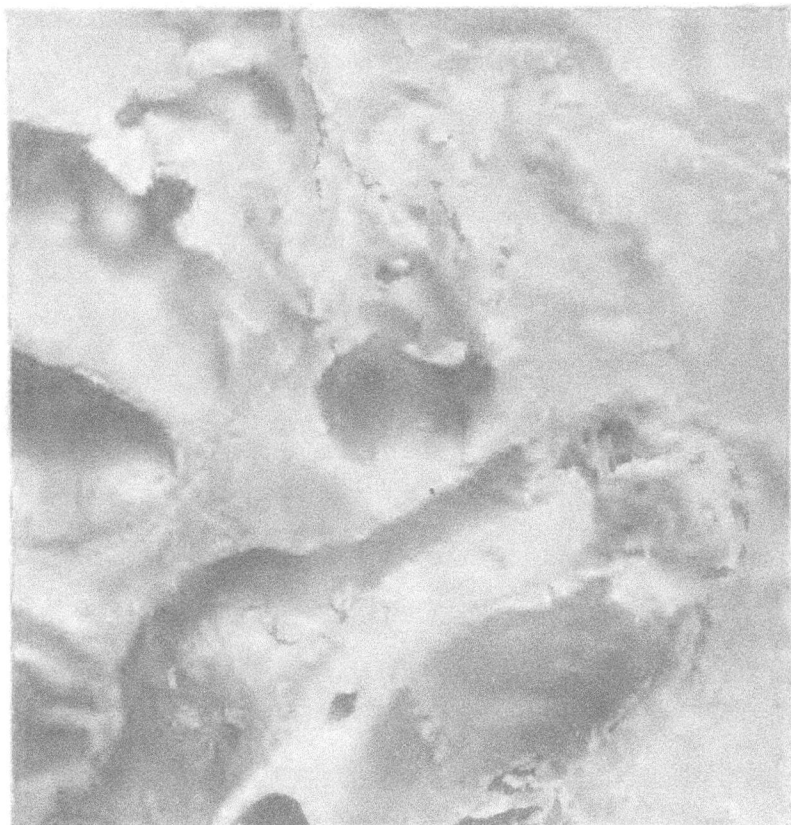

FIG. 4.

opening to investigation the cavity holding the belly of the stapedius.

That this, my concluding report on the specimen under discussion, should have presented but little additional information to members of the Society, I sincerely regret. If, however, the latter part, containing the four new illustrations, serve to make clearer what was attempted to be set forth in the two papers already given the Society by Dr. Buck and myself, and what all previous illustrations have, in the opinion of both writers, but imperfectly shown, I shall myself feel in some measure consoled for the shortcomings of its former part, and shall hope that members of the Society will feel that the paper, having subserved a useful purpose, has not been presented wholly in vain.